T0146030

RAND

The Potential of Nanotechnology for Molecular Manufacturing

Max Nelson, Calvin Shipbaugh

ISBN: 0-8330-2287-3

RAND is a nonprofit institution that helps improve policy and decisionmaking through research and analysis. RAND® is a registered trademark. RAND's publications do not necessarily reflect the opinions or policies of its research sponsors.

© Copyright 1995 RAND

All rights reserved. No part of this book may be reproduced in any form by any electronic or mechanical means (including photocopying, recording, or information storage and retrieval) without permission in writing from RAND.

Published 1995 by RAND
1700 Main Street, P.O. Box 2138, Santa Monica, CA 90407-2138
1200 South Hayes Street, Arlington, VA 22202-5050
201 North Craig Street, Suite 202, Pittsburgh, PA 15213-1516
RAND URL: http://www.rand.org/
To order RAND documents or to obtain additional information, contact Distribution Services: Telephone: (310) 451-7002; Fax: (310) 451-6915; Email: order@rand.org

Preface

This research was undertaken to explore the potential for advanced manufacturing based on molecular nanotechnology. This report provides a framework for understanding the scope of this topic—possible benefits, development risks, and policy options—but it is not the intention of the authors to provide a definitive road map (with corresponding technical assessments); nor is it believed by the authors that such a detailed analysis would at present yield a fully credible road map. Rather, it is the contention of the authors that much basic and applied research needs to be undertaken to realistically assess the far-term viability of many of the most interesting emerging concepts, but a careful and objective feasibility assessment could help stimulate near-term achievements and prevent technological surprise by foreign players. The authors anticipate that the framework and analysis presented herein could provide useful and objective input into such a technology assessment.

RAND supported this research as part of its corporate-sponsored research program. This report should be of interest to policymakers, scientists, and other individuals involved in the fields of nanotechnology and molecular manufacturing.

Contents

Figures

Summary

Nanotechnology—a term introduced in 1974 to describe ultrafine machining of matter—has come to be applied to a wide scope of small-scale engineering. With nanotechnologies, two activities are possible—nanomeasurement and nanomanipulation. Molecular manufacturing is the willful use of these two activities to create objects. Proponents of the application of nanotechnology to molecular manufacturing suggest that environmentally clean, inexpensive, and efficient manufacturing of structures, devices, and "smart" products based on the flexible control of architectures and processes at an atomic or molecular scale of precision may be feasible in the near future (i.e., 10–20 years from the present). The ambitious goal is to produce complex products on demand using simple raw materials; e.g., inserting the basic chemical elements in a molecular assembly factory to yield a common household appliance, perhaps with sensors and actuators built-in to respond to commands or environmental conditions. The question of whether it is possible to achieve a stage in the foreseeable future when such extreme capability might be viable, and if so how to develop the field, is a point of contention in both scientific and policy circles.

The concept of manufacturing at the "nano" or atomic scale dates to more than three decades ago. Many developments in biotechnology, chemistry, computational tool building, electrical engineering, and physics have moved the scientific and engineering community closer to operating smoothly on the nanoscale. In addition to extensions of micromachining—with production methods such as lithography, commonly encountered for microelectronics or microelectromechanical systems (MEMS)—there have been recent developments in scanning force microscopes (SFM), using probes that can position atoms or molecules to nanometer scales, and interest in investigating the means by which complicated molecules with desired properties can be modeled, synthesized, and perhaps even self-assembled. These recent developments have motivated advocates of a "bottom-up" approach for manufacturing molecule-by-molecule.

Exclusive use of this approach however, misses the longer-lived history and some of the benefits being achieved through the more familiar "top-down" approaches. The top-down approach is one in which macroscale components are utilized to create nanoscale structures. This differs from the bottom-up approach, which uses nanoscale components to create structures. In particular, top-down structures and methods might help with the interfacing of bottom-up structures into a system. Cases to support this position include chemical sensors that use microelectronics technology, biosensors that use enzymes and electrodes, and the potential of protein-based memory in an optical holography system.

Useful means of positioning and interconnecting molecular structures might be created in the near term that could serve as a proof-of-principle that more ambitious molecular manufacturing may be possible. If meaningful molecular assembly (or more extensive modeling tools for rational molecular design) is not demonstrated in the next decade, then the field of molecular nanotechnology may well have encountered an impasse that will challenge the credibility of the practicality of molecular nanotechnology for a revolution in manufacturing concepts.

Extensive molecular manufacturing applications, if they become cost-effective, will probably not occur until well into the far term. However, some products benefiting from research into molecular manufacturing may be developed in the near term. As initial nanomachining, novel chemistry, and protein engineering (or other biotechnologies) are refined, initial products will likely focus on those that substitute for existing high-cost, lower-efficiency products. Likely candidates for these technologies include a wide variety of sensor applications; tailored biomedical products including diagnostics and therapeutics; extremely capable computing and storage products; and unique, tailored materials (i.e., smart materials using nanoscale sensors, actuators, and perhaps controller elements) for aerospace or similar high-cost/high-capability needs. The current development of MEMS devices may open avenues for incorporating molecular nanotechnological components into widely used systems, such as automotive parts.

As indicated by the large number of U.S. research centers involved in molecular manufacturing and nanotechnology, the United States is a leader in this field. The majority of these centers are in academia and often consist of a few investigators in one or two departments. Identified activity represents a diverse set of basic or applied studies in materials properties. This is vital to providing the building blocks for a technology development. However, this is being done largely without a plan representing an organized, embracing systems-development goal of molecular nanotechnology.

A key observation is that a number of countries are engaged in some level of effort relevant to the foundations of molecular nanotechnology. Although the United States has many groups performing work related to nanotechnology and molecular manufacturing, there are several strong competitors and potential collaborators. Japan has large efforts that are funded individually at a significantly higher rate than their U.S. counterparts and are coordinated by a dedicated national effort. Other nations with strong research centers include China, Denmark, France, Germany, Russia, Sweden, and the United Kingdom.

It is unclear which fabrication method will best succeed—multiple research paths should be left open at the basic and applied research level. Areas that are important to the future of molecular nanotechnology-based advanced manufacturing, and in which successful discoveries could serve other applications in the interim, include the following:

- Macromolecular design and folding
- Self-assembly methods
- Catalysis (inorganic, enzyme, and other)
- Dendrimers, fullerenes, and other novel chemical structures
- Bioenergetics, nanobatteries, and ultrasound-driven chemistry
- Semiconductor-organic/biological interfaces
- Miniaturization and massive parallelism of SFM
- Molecular modeling tools.

The potential is enormous and could lead to extreme miniaturization in space systems, capabilities in human performance enhancement and medical treatment, as well as ability to manufacture a wide variety of sophisticated products on demand. It might be expected that if sufficient applied science checkpoints are passed, then manufacturers would be motivated to pursue development of applications.

Past experience with translating science into practical engineering provides cautionary examples as well as successes. In principle, civilization can make use of controlled nuclear fusion as an immense source of energy in analogy with nature's application of various fusion reactions to power stars. However, the reality of achieving this has been much more difficult than originally anticipated. Similarly, achieving the manufacture and control of sophisticated molecular nanodevices from current conceptual designs may be more difficult than anticipated.

A fully credible assessment of how far molecular manufacturing will progress in the next two decades is not possible until incremental steps have been undertaken, although tentative indications appear positive. At present, modeling and theoretical underpinnings need to be further developed.

Demonstration of assembly, control of chemistry, and practical component creation and integration are important. The laboratory development of several steps should be closely followed for indications that milestones can be expected:

- Produce material parts at the nanoscale.
- Process material parts into components at the nanoscale.
- Order molecular components into structure and interconnect.
- Interface system components with the macroenvironment.
- Control a massive collection of miniature parts and systems.
- Provide a power system.

The many laboratory steps needed indicate that a careful decision on development policy, if any, should be made. There are several options:

- Maintain a laissez-faire policy toward coordination of research efforts and resources.
- Conduct a detailed, objective working group technology assessment of the state of the art of relevant molecular nanotechnology research and potential applications.
- Establish a coordinator or program to oversee research and developments.
- Create a national or international cooperative effort.

To prevent the possibility of technological surprise, yet not prematurely enact policies that commit funds and valuable resources, a prudent course of action would be to create a working group of biotechnology experts, chemists, computer scientists, electrical engineers, materials scientists, mechanical engineers, and physicists. This group's assessment of a laissez-faire posture versus coordination and cooperation should then be implemented as a basis for a rational policy about support for molecular nanotechnology.

Although there has been much encouraging theoretical and conceptual study of the advanced manufacturing potential of molecular nanotechnology (and panel reports and surveys of expert opinions), a comprehensive, detailed technical assessment by a multidisciplinary, objective expert working group is lacking and should be conducted to determine engineering feasibility. The role of ultra*fast* phenomena in manufacturing methods and the issue of what applications these could address can be included in conjunction with an assessment of nanoscale technologies (as a secondary focus of how to exploit extreme-scale phenomena).

A positive finding from such an assessment would indicate that cooperation at the basic and applied research level beyond the present situation should be organized. Increased coordination of research funds may improve the cost-effectiveness by reducing redundancy; however, such increased organization should be done in incremental steps so that it does not come at the expense of healthy competition. A negative finding from such an assessment—such as low engineering feasibility; low potential for viable, near-term application; or limited prospects for critical research progress—would strongly indicate that the current levels of funding and structures for basic scientific research in molecular-based nanotechnology is appropriate and that extensive resources should not be dedicated to developing specific pathways.

Acknowledgments

The authors wish to thank the formal reviewers Randall Steeb and Howell Yee for their constructive comments. The authors would further like to thank Tom McKendree for useful insights and information. We would also like to express great appreciation to Paul Davis for providing support and assistance.

1. Introduction

Definition

The term nanotechnology was introduced by Taniguchi in 1974 to cover machining in the 0.1 to 100 nanometer range.[1] This definition has been extended to include methods based on chemical synthesis and biotechnology. In addition to extensions of micromachining—with methods commonly encountered in microelectronics such as lithography—there are more recent developments in scanning force microscopes using probes that can position atoms or molecules to nanometer scales, and interest in investigating the means by which complicated molecules with desired properties can be designed, synthesized, and self-assembled. These more recent developments constitute molecular nanotechnology.

Motivation

With nanotechnologies, two activities are possible—nanomeasurement and nanomanipulation. Molecular manufacturing is the willful use of these two activities to create objects. Proponents of the use of nanotechnology for molecular manufacturing suggest that environmentally clean, inexpensive, and efficient manufacturing of structures, devices, and "smart" products based on the flexible control of material architectures (and processes) at an atomic or molecular scale of precision may be feasible in the near future. Although there may be concurrent use of many different manufacturing methods to create miniature components and features, they present distinct engineering routes to fabrication of materials and devices. We will discuss molecular manufacturing, which is based on methods from chemistry, biology, and several engineering disciplines. Molecular manufacturing should not be confused with micromachining—the latter is a top-down approach that attempts to use macroscale components to create nanoscale structures, while molecular manufacturing proposes a bottom-up approach to build systems and devices from the atomic or molecular level.

---

[1]The prefix nano refers to a billionth part (10^{-9}) of a meter.

The question of how to develop molecular nanotechnology to a point where a wide variety of significant applications might be achieved—and what goals are realistic—is a point of contention in both scientific and policy circles. Synthesis and exploratory work by K. Eric Drexler, among others, have suggested that molecular manufacturing has the potential to enable commercial production of superior products in the near future.[2]

This report describes some potential risks and benefits of investment in molecular manufacturing and attempts to provide policymakers and the public with information that should be helpful for understanding the nature of molecular nanotechnology developments and assessing appropriate support and organization.

_____

[2]Concept exploration can be found, for example, in the popular books, Drexler, 1986, and Drexler, Peterson, with Pergamit, 1991.

2. Trends and Goals

Historical Developments

Many of the concepts of molecular-based nanotechnology predate the previously cited synthesis and exploratory works of K. Eric Drexler (and colleagues).[1] The concept of manufacturing at the "nano" or atomic scale dates to more than three decades ago when Nobel laureate physicist Richard Feynman gave a lecture entitled, "There's plenty of room at the bottom."[2] Feynman described a vision for a world with technology that could etch lines a few atoms wide with beams of electrons, build circuits on the scale of angstroms to create new computing structures, and manipulate atoms to control the very property of matter. Since Feynman's landmark 1960 lecture, many developments in chemistry, electronics, and tool building have moved the scientific and engineering community closer to operating on the nanoscale (although the molecular manipulation methods envisioned today are often different from those in Feynman's lecture).[3] Figure 1 presents several important developments.

An early demonstration of the potential for molecular manufacturing occurred in 1920 with the development of the Langmuir-Blodgett (LB) thin film production technique. This is a molecular deposition method that allows the formation of ultrathin film, monolayer (and subsequent multilayer) structures that can be used to order the arrangement of molecules and, in particular, have other types of molecules embedded in a layer for functional purposes. It is often suggested that this can serve in the near future as a tool for assembling many types of molecular components of a system. One example given is an artificial biological machine that could use an artificial photosynthetic center composed of LB layers to manufacture useful organic molecules (e.g., fuel) (Gopel and Ch. Ziegler, 1992).

Modern proponents of molecular manufacturing often focus on a bottom-up approach, advocating research and development efforts in such areas as molecular synthesis and scanning microscopy. Exclusive use of this approach

---

[1]Many contemporary technical arguments for molecular nanotechnology can be found in Drexler, 1992.

[2]An address by Richard Feynman (Feynman, 1960) is often taken to be a seminal focus in the history of nanotechnology.

[3]Examples of interesting developments can be found in Whitesides et al., 1991, and Schneiker, 1989.

4

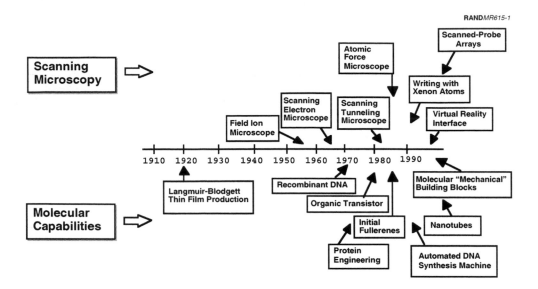

Figure 1—Past Developments Toward Molecular Nanotechnology Capability

however, misses the longer-lived history and some of the benefits being achieved through the top-down approach. In particular, top-down structures and methods might help to interface bottom-up structures into a system. One example of this would be support for the creation of advanced semiconductor-organic (including biological) interfaces that might be a step in the direction of providing novel, workable computer architectures (Kaminuma and Matsumoto, 1991). Other cases in point are chemical sensors using microelectronics technology, biosensors that use enzymes, and the employment of protein-based memory in an optical system (Birge, 1992). Although these examples may represent macroscopic-scale systems, they employ nanoscale materials and thus may properly be considered relevant applications derived from molecular manufacturing.

Top-down nanomaterials processing benefited from the demands of the microelectronics revolution—for example, the development of evaporation and condensation processes (around 1974), or plasma processes in 1981 to produce ultrafine powders (Franks, 1991). The top-down approach also led to early advances in nanomachining. For example, ion beam machining developed around 1967, showered ion beam machining developed around 1978, and ultra-fine turning and grinding technologies developed around 1983.

Although LB techniques offered early potential for some development, it was not until the appearance of the scanning tunneling microscope (STM) in 1982—and subsequent development of probe tools—that bottom-up nanomachining and nanomaterials processing became a clear possibility, as evidenced by the ability to draw maps and write names with individual atoms or small globs of atoms through the force such probes can exert on surfaces.

Recent developments in chemical synthesis, biotechnology, and molecular modeling might open new pathways to rationally construct products from the bottom-up. For example, chemists have worked with colloids and suspensions for a long time. Within the last 20 years, a very active area of research has been the solution chemistry and physics of *quantum* particles (with sizes under 50 angstroms). Both metallic particles and semiconductor particles have been synthesized and used in novel ways.

In other developments in recent years, there has been progress in exploration and chemical synthesis of nanoscale structures such as fullerenes and electronic components. Research on organic conductivity dates back to the 1950s; doped polyacetylene film conductivity was demonstrated in 1977; and the creation of organic-molecular active components proceeded in the 1980s with various labs having reported the production of molecular wires, switches, diodes, and even transistors (e.g., thin film organic semiconductor layered devices). These are components that would make molecular electronic computation possible, if massive assembly were practical. Although further developments along these lines may provide for the suitable manufacture of electronic components, the architecture needs to be thought through and the speed of the components optimized. At present, if a molecular computer could be fashioned with these components, it would likely not compete with a silicon-based digital computer even though specialized circuits and applications, such as in displays, may be enticing.[4]

However, there are systems that currently use molecular electronic components to good advantage. A number of types of molecules have been used in holographic memories; i.e., making use of nonlinear optical materials. Excitation times can drop considerably below a nanosecond (allowing rapid switching times in computational architectures), and memory could theoretically approach a trillion bits per cubic centimeter.

In addition to the emergence of molecular electronic components, there is an emerging potential for molecular mechanical systems as well. Rods and rings have been structured from molecular chains such as staffanes and propellanes. Interlocking structures (not chemically bonded) have been reported (Kaszynski

---

[4]One recent invention is the symetrically configured alternating-current light-emitting (SCALE) device by Arthur Epstein and colleagues. An advantage is that the fabrication of such thin film devices is made easier and presumably cheaper with polymer processing than is the case when using only inorganic layers (Dagani, 1995).

et al., 1992). This raises the possibility of obtaining variations in properties by adding functional groups to existing molecules. Issues include whether these structures can be made sufficiently stable and manipulated in a practical fashion with an appropriate "engine," including transducers and a power source. K. Eric Drexler has suggested molecular machinery built around diamondoid structures, which might be very hard and provide extensive building blocks if, for instance, closing planes of such material into stable cylinders can be suitably directed. Even if this example of very hard structures is not viable, there remains a motivation to model and explore the extent to which molecular structures can be mechanically exploited.

Moving from positional and synthetic chemistry paths to conceivable biological paths for molecular manufacturing, the tools include genetic engineering (e.g., recombinant DNA and DNA synthesis machines) to create proteins and DNA amplification such as by polymerase chain reaction (PCR). The latter is capable of multiplying a microscopic sample of DNA in a matter of hours to macroscopic proportions using simple reagents and a small number of heating-cooling cycles that allow bonds to break and reform. This is a reminder that it is possible to rapidly produce molecular "information-template" components. What is needed is an assembly process to join components into systems.

In addition to using the ultraminiaturization of spatial scales as an organizing structure for technological applications, the exploitation of ultrafast phenomena might be considered. Ultrafast phenomena are those that occur on extremely short time scales, such as pico- or even femtoseconds (10^{-12} or 10^{-15} seconds, respectively). Laboratory developments in recent years in pulse-shaping optical equipment and generation devices have made the study of such extreme phenomena a reality. Early work indicated a potential for bond-selection in molecules (i.e., subnanometer scale effects) to drive the direction of simple chemical reactions by lasers (Moffat, 1992). The vibrational modes of microtubules in cells are another example of phenomena with extremely short time periods that in principle could be used (e.g., as a biomolecular switching function) (Hameroff et al., 1992). Ultrafast technologies may serve in conjunction with, or separately from, nanotechnology.

Though still a nascent field and far from producing commercial products, significant laboratory advances have been made in bottom-up molecular manufacturing. A number of fundamental, frequently independent, breakthroughs in recent years may indicate a synergistic growth is possible for molecular manufacturing.

Potential Benefits

Near Term

We have briefly reviewed a number of the research developments that give credibility to the pursuit of extreme control and flexibility in molecular manufacturing. The question arises, "What specific application areas might these developments impact in the next one to two decades?"

As early nanomachining and protein engineering technologies are developed, initial products will likely focus on those that substitute for existing high-cost, lower-efficiency products. A number of concepts in microelectromechanical systems (MEMS) are in the process of rapidly emerging as contemporary applications (Brendley and Steeb, 1993). Speculatively, extrapolating from the success of microelectronics and MEMS devices, we might anticipate that molecular nanotechnology, if engineering routes are demonstrated feasible, could contribute to such applications as: significant advances in categorization, or understanding at the atomic scale (e.g., fundamental exploration of frictional forces), and subsequent manufacturing of new advanced (lightweight) materials for strength or heat-resistance (changing defects or adding chemical "dopants" can enhance or degrade materials to create unusual properties); the ability to manufacture markedly enhanced semiconductors for use in more conventional microtechnologies (e.g., control dopants at a very small scale); high-density, rapid computer storage systems; massively parallel computing based on molecular or biomolecular functions and integration; improved sensing technologies that could be used in a wide range of industrial quality control and repair systems (e.g., aeronautic or other transportation applications to provide not only warnings and diagnostics but also adjustments en route); detection and identification of chemical or biological weapon threats through specificity of molecular response; improved gene sequencing; and medical diagnostics and therapy (e.g., synthesis of artificial muscles and many other tissues—some synthetic materials have already been investigated with biotechnology).

As an offspring (in part) of biotechnology, molecular manufacturing may indeed offer promise in aiding human health and performance. On these grounds alone, efforts in molecular nanotechnology could be well worth pursuing. An example that may seem exotic when one first considers it, but may be possible given current laboratory research, is the often cited emergence of a miniature "submarine" that might detect problems and even perform operations within the circulatory system. This concept predates Feynman's seminal work and might have a chance of being realized in the not-too-distant future with a vigorous research and development program combining various developments of biotechnology and nanotechnology. In a different application, the rapid development of gene sequencing capability may be combined with design of molecular machines to give strong new potential in gene therapy (gene sequencing and splicing in the agricultural arena has been shown to have high value in such applications as deriving antibiotics or imbuing crops with resistance to pests). Yet another medical application that could result from molecular nanotechnology is a "smart" pill that senses the chemistry of its body environment and responds to health conditions by releasing measured doses of the appropriate drug.

Far Term

Extensive molecular manufacturing that would allow the positioning of atoms and molecules as desired into almost any conceivable pattern at a macroscopic scale is the ultimate goal of the most ambitious forms of molecular manufacturing. Related concepts have been a theme in science fiction stories.[5] Early advocacy of molecular nanotechnology has been greeted with skepticism from strong critics and has even included a statement that powerful advocacy could constitute a "nanoreligion." It can currently be argued that a level assessment would not find that "molecular nanotechnology" (whatever that specifically means to the user) is restricted to the substance of science fiction, but postulated revolutionary manufacturing and application technology goals must be carefully assessed, and development must be pursued in a calculated fashion. If the more extreme nanoscale manufacturing capabilities become possible, it will probably not occur until well into the far term.

Many concepts have been forwarded that take the idea of controlling activity at the atomic and molecular scale to an extreme. In analogy with the current situation of studying and applying naturally occurring microbes to assist in cleaning up the environment (such as degradation and break-up of offshore oil spills), nanoscale products and devices have been proposed as someday being applicable to the remediation of hazardous waste sites containing multiple contaminants, the removal of toxic products in manufacturing industries, and even the opening of rapid pathways for the terraforming of other planets. Production of miniature machines that reside in the body as sentry units to prevent disease-causing organisms and agents from prospering, or other nanomachines to repair physical damage to tissue have been offered as a vision

---

[5]The idea of using manipulative hands to create smaller manipulative hands, and to combine this with a scanning observation, to perform miniature scale surgery was introduced in fiction by Heinlein, 1940.

for ultimate health maintenance and control of the aging process. The manufacturing of industrial or even household objects that can sense and react to circumstances (or a master's presence) is suggested for smart structures that border on the functional appearance of "sentience"; e.g., furnishings that adjust shape, color, and texture to conform to different styles when commanded, and even anticipate moods from body language or clothing.

Many of the discussions advocating or disputing such advanced molecular nanotechnology-based advanced manufacturing have been overly focused on debating fundamental scientific laws at one extreme, and in other cases debate has revolved around select exploratory concepts. An argument that has been given to counter concerns about whether basic laws of science prohibit molecular nanotechnology appears to be rooted in a biological "proof of existence." This argument essentially cites the origin of life as evidence for sophisticated molecular machines. While nature has indeed created very sophisticated molecular machines, it is not evident that mankind will be able to achieve in practice what he could in principle.[6]

A fully credible assessment of how far molecular manufacturing will progress in the next two decades is not possible until several steps that would be expected along candidate pathways have been demonstrated, although tentative indications are apparent. At present, modeling and theoretical underpinnings need to be further developed. A development path toward a smart molecular manufacturing capability should also, if frustrated from reaching its ultimate goal, produce many intermediate and fruitful successes; i.e., plan a program that can have a successful outcome at termination even if the primary goal is not met. Should laboratory breakthroughs occur, the potential is enormous and could eventually lead to an ability to manufacture a wide variety of structures and devices (for many disparate applications) on demand; e.g., extreme miniaturization of space exploration systems and multiplication of small devices to take advantage of indigenous resources found on asteroids, comets, or planets for mining; defending Earth against impacts; or tools to assist extensive colonization of the solar system on a reasonable time scale.

---

[6]The sun generates energy based on nuclear fusion. In principle, civilization can make use of controlled nuclear fusion as an immense source of energy. However, the engineering reality of achieving this (other than with peaceful uses of thermonuclear explosives) has been much more difficult than originally anticipated. Similarly, achieving sophisticated molecular machines may be more difficult than anticipated from conceptual designs.

3. Developing Incremental Checkpoints

Elements of Nanofabrication

Individual STM probe tips can be built macroscopically, but the needs of massive parallelism and rapid rate manipulation for manufacturing purposes at the nanoscale push device requirements in the direction of using very miniature probe tip arrays. One of the problems that has been pointed out for the creation of a molecular nanotechnology-based advanced manufacturing industry is that the very tools that can function to move extensive numbers of atoms or molecules around in a rapid fashion must themselves be built from nano- or micro- instruments.[1] Practical engineering routes for additional miniaturization must be confirmed.

The several alternative approaches for molecular manufacturing must be separated or their overlaps and possible synergisms identified. The next several paragraphs will discuss the block components of Figure 2.

The fabrication of structures at the nanoscale is done by nanomachining and nanomaterials production methods based on molecular engineering techniques. Each set of fabrications is guided by a control process employing sensing information provided by nanomeasurement methods. Nanomeasurement is the observation of nanoscale entities and phenomena; e.g., the characterization of an LB film by a scanning force microscope (SFM).

Nanomachining can be defined as the fabrication of nanoscale structures with tools that remove or add material on a base (e.g., silicon), similar to adding bricks or digging holes to construct a building. Nanolithography is a set of top-down methods that have been employed with great success. Other machining methods include the deposition or growth of layers (by any of several techniques), and abrasive ultraprecision finishing and forming of surfaces.

The U.S. National Nanofabrication Facility has routinely created useful laboratory tools and products with these techniques. As the applications of microelectromechanical systems multiply, the ability to perform top-down

---

[1]Ball, 1993, performs a book review of Drexler, 1992, and Crandell and Lewis, 1992, and discusses the aims and meaning of nanotechnology in various contexts.

RAND*MR615-2*

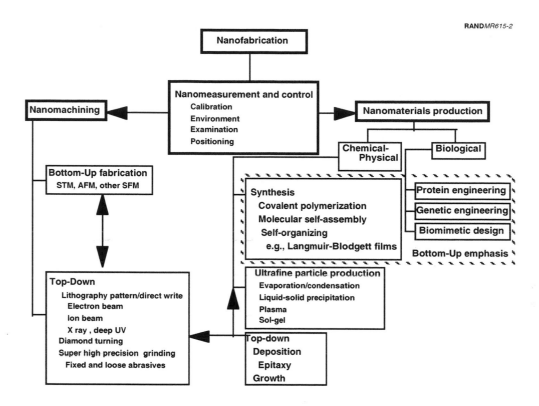

Figure 2—A Taxonomy for Understanding Nanofabrication

nanomachining will become a valuable resource to augment the familiar regimes of micromachining.

An SFM uses miniature probes that can sense the surface of a material (to a resolution that is sometimes as small as the atomic scale) through one of many mechanisms; e.g., a magnetic force microscope could use a magnetized probe to study magnetic media. These probes function via a feedback mechanism. An STM maintains a constant tunneling current by providing feedback to a piezoelectric device. The STM can in turn be used as a measuring tool for an atomic force microscope (AFM). An AFM can probe nonconducting surfaces using a tip on a lever arm deflected by the atomic forces encountered at the surface of the sample being studied.[2] The tools used in top-down micromachining might be used to generate various SFM tools that have been suggested for bottom-up atomic and molecular structure building.

Molecular materials can be fashioned by biological methods, chemical synthesis, or a combination of chemical and physical processes. Biomimetic design is the

---

[2]An alternative is to use a laser to measure AFM deflections.

use of information gained from studying living structures and processes to create artificial products, exploiting what evolution has developed and perhaps improving upon it. For example, the study of the nanoscale structure of abalone has led to a composite material, B_4C/Al, with very desirable mechanical properties (i.e., high fracture toughness and strength) (Cahn, 1990).

Bridges between top-down and bottom-up thinking exist. Tools that deposit or grow materials are one such case. Deposition is used in micro- or nano-machining to give very thin films. Molecular beam epitaxy is a method that can generate extremely thin layers of one atom thickness at a time and is an integral part of processes that use top-down techniques to generate quantum structures from alternating layers of material to produce two-dimensional quantum wells, one dimensional "wires," and quantum "dots" that can be used in optoelectronic parts such as efficient miniature lasers and switching devices.

Key Steps

Two general alternative methods—scanning force microscopy and macromolecular design and fabrication—illustrated in Figures 3 and 4 show some major stepping stones toward "smart" molecular manufacturing. Many proponents of nanotechnology focus on this end point. There is no clear point in time to expect such capability, but as shown in the figures there are many significant applications that should be expected to arise on the path toward smart molecular manufacturing. Before reaching any of these applications, there are a number of fundamental tool developments that must be demonstrated, such as the development of miniaturized, high-rate scanning probes and precision control technologies, perhaps including virtual reality interfaces to assist operators.

Nanostructure fabrication using tools such as the STM and the AFM has been offered as a positional chemistry in which desired reactions between precisely defined and aligned molecules are controlled (distinctions exist between whether the SFM tip reactants are delivered under vacuum or in solution) (Drexler, 1992). The benefits from traditional nonpositional chemistry, which relies on random collision and suffers from numerous unwanted reactions, are considerable, and it might be imagined that the deliberate control of the direction of chemical reactions will introduce additional capabilities. As Figure 3 indicates, chemical or biological synthesis may work together with probes in the "assembler." One conception of the assembler as a tool for nanostructure fabrication that has been suggested in the literature is a device having a submicroscopic robotic arm(s) (i.e., SFM probe tips) under computer control capable of holding and positioning

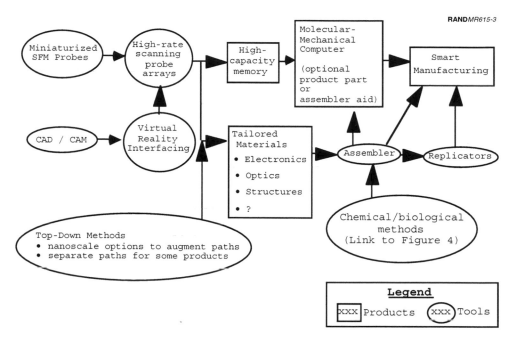

RAND*MR615-3*

NOTE: CAD / CAM = computer-aided design/computer-aided manufacturing.

Figure 3—Steps Toward Molecular Manufacturing with Scanning Probes

reactive compounds, with respect to molecular workplaces and devices, to control the precise location at which chemical reactions take place (Drexler, 1992).

The development of atomic force probes is well along, and even high school students are able to do some form of high-resolution scanning tunneling microscopy. In fact, the tools for producing crude forms of STMs have probably been in existence since the 1960s.

It has been suggested that a series of increasingly complex demonstrations of carrying through a chemical reaction under the control of a probe is a natural step to anticipate in the near future. Such results could be encouraging and might lead to the first simple assemblers within the next several years. One encouraging line of research has reported that the next several years may see a laboratory result that resembles an assembler.[3] The approach is to remove hydrogen atoms with a probe, leaving reactive sites on a surface, and then allow chemical catalysis to assemble molecular structures. It was suggested that the

---

[3]This was described in a presentation by William Goddard III, California Institute of Technology, at the Third Annual Foresight Conference on Molecular Nanotechnology, October 1993, Palo Alto, CA.

process be terminated by adding the proper chemical solution, or by bringing forth a terminating surface.

The success of this type of experiment is an example of a milestone that might be sought as indication to proceed further with support for molecular engineering studies aimed at developing a versatile manufacturing capability. Failure to succeed in establishing some versatility in laboratory chemistry with probes in the next several years could indicate this approach to molecular manufacturing is risky.

For an assembler to become practical, the ability to be able to position atoms or molecules at a high rate is the type of breakthrough required. There has been progress in this direction, for instance by using parallel arrays of probes. This can in principle lead to tremendous semiconductor storage densities (terabit chips) by moving atoms (bits of information) with an array of tips (MacDonald, 1992). These microscopic probes are at present much larger than examples (e.g., around 100 nanometers) described by Drexler for robotic arms in his proposed assembler (Drexler, 1992). If the challenges to making such very small arms can be met, the benefits in speed are potentially that much greater and could, in principle, exceed holographic storage densities.

Introducing mechanical computational circuits will first require some demonstrated ability to manufacture and assemble movable parts (i.e., appropriate materials). This is independent of realizing success in digital storage by manipulating atomic positions. The laboratory proof of miniaturized component production (probably dependent on appropriate chemical/biological tools), stability, and interconnection ability are needed, as well as further study of issues surrounding reversible operations and limits of computation, as discussed by Drexler, Merkle and others (Crandall and Lewis, 1992). If successful, the small energies per operation and density of circuits would be outstanding. The production of a simple circuit example does not in itself, however, prove the eventual ability to produce an extensive system of computational circuits.

Analogous to robotics, nanosystems (sensor, memory and processing, actuator, power, or communication) might be organized into preprogrammed effectors that carry out a sequence of tasks, teleoperated devices controlled through an interface (chemical or otherwise) with larger systems, or fully autonomous robotic devices. John von Neumann introduced a theory of replicating automata, and various mechanisms have been suggested to speed up molecular manufacture via the fabrication of replicators, which would carry the instructions to make copies of themselves much like cellular reproduction. This seems like a relatively risky route to manufacturing, even though nature has succeeded with DNA programmed instructions (and RNA in retroviruses).

Figure 4 is a flowchart for the use of chemical synthesis or biological techniques. For either of these approaches, molecular modeling developments might be expected to play a vital role. As indicated earlier, it is felt that this approach benefits from interaction with probe tool development. The end point, once again, is smart molecular manufacturing, with assemblers and replicators being important concepts that require demonstration.

The chemical synthesis route might be based on any of many interesting developments (Whitesides et al., 1991). Fullerenes, catalytic antibodies, and various self-assembling/self-organizing polymers and films might augment or bypass the assemblers and replicators indicated in Figure 3.

There are some differences in products in Figure 4 from those indicated in Figure 3 that are not apparent. High-density memory for this example is already demonstrated by holographic memories in a more mature way than the exciting (but currently impractical) atomic positioning that has been shown with probes. The latter is a transient laboratory phenomenon and the former a rather practical

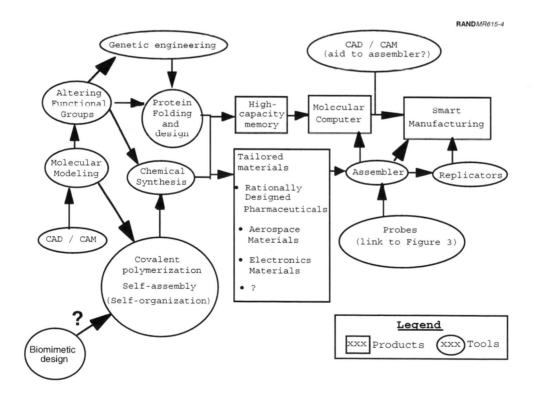

Figure 4—Steps Toward Manufacturing with Chemical-Biological Techniques

accomplishment already. The issue is whether chemical and biological methods will yield further practical contributions to memory storage. Pharmaceuticals are more directly related to the lines of approach in Figure 4 than to the probe path of Figure 3.

Genetic engineering is a well-established area, and it is really the design of improvements to proteins that is of chief interest as a tool for building useful biological materials. In nature, proteins serve structural and other functions—e.g., response to photons.

Proteins offer a natural technology in use since the beginning of life itself. Given the suspected influence of prions on organisms, even small protein-like structures can offer the potential for control of function in some complementary molecular process. Protein and protein-like polymers are created by stringing together amino acids or other raw material in a chain. Protein production in cell ribosomes is an example of a molecular manufacturing technology. Typically, a ribosome is on the scale of a few thousand cubic nanometers and builds various proteins by bonding amino acids together in a precise sequence under instructions provided by a messenger RNA. RNA can be thought of as the control program for the factory, and ribosomal RNA exists as "workbenches" for the processing (Hall, 1993). Utilizing natural molecular manufacturers, significant benefits have already been realized in molecular biotechnology. Further development could lead to medical benefits as well as benefits in materials science and chemistry and electronics and computing.

Protein engineering is not the only subject for molecular assembly. Other polymers are also candidates, and chemical synthesis seems like a promising area to find building blocks for molecular machines. A prime consideration for assembly and molecular design is the development of molecular modeling aids. This is an area that in the near term might have high leverage if supported. Should the folding problem prove intractable after several years, support for this approach might have to be reconsidered, but not necessarily dropped since improvements in this area would also have benefits in, for instance, medicine through molecular design improvements.

In chemistry and biology, catalysis has a special role. The advantage of catalysis is that the "machine" that organizes reactants is not permanently altered or consumed, and the nature of catalysis allows reactions to proceed at appreciable rates using reasonable temperatures and pressures. Catalysts have found use in precise molecular manufacturing already—for example, in the control of stereoregularity among polymers. Enzymes are protein catalysts. The thermal nature of satisfying the "lock-and-key" model of enzyme catalysis means that individual steps are relatively slow, implying sufficient time be allowed, or quantity of catalyst used, for massive manufacturing. Enzymes and inorganic catalysts can play a role in directing synthesis given that the proper catalyst is found—catalysis tends to be a specific process. One recent area of research, catalytic antibodies, offers potential for construction in molecular manufacturing.

Dendrimers are a novel class of nanoscale molecules that grow in predictable patterns into massive molecular structures. The particular structure (and hence properties) can be tailored and might be forced to exhibit some biomimetic property (e.g., gene therapy might be an application). Catalytic and other functions can be multiply distributed on dendrimers—they might be used as sensors as well as fabricators (Tomalia, 1995).

Developments in CAD/CAM are important to molecular manufacturing. Graphics packages that allow visualization of molecular structures, calculations that refine the physics predictions of molecular states, and specialized routines such as those that consider and simplify the combinatoric problem of polymer folding are all exemplar candidates for areas that can aid in rational molecular design. The United States seems well situated in terms of both software and computational power with respect to the world. Display development is an important area to invest in to get the greatest return in modeling and control. Recent work at the University of North Carolina in cooperation with other institutions has shown remarkable progress in connecting an STM with a virtual reality system that allowed the user to "feel" his way over a microscopic surface.[4]

Development Risks

There are no laws of nature that are clearly being violated by the general concept of molecular nanotechnology-based manufacturing.[5] Roadblocks may arise, however, from any number of sources—lagging developments in applied science, model development, funding, or the engineering practicality of theoretical concepts.

The development of a proto-assembler as a milestone (key checkpoint) for government support has been recognized (OTA, 1991). More detailed

---

[4]Russel Taylor of the University of North Carolina presented this research at the Third Annual Foresight Conference on Molecular Nanotechnology, October 1993, Palo Alto, CA.

[5]Particular concepts may be offered that are found to have flaws in basic scientific feasibility, so an examination of compliance with the laws of nature is a necessary first step for any idea proposed in molecular manufacturing. However, such flaws should not be generalized as applying to the entire concept of molecular manufacturing (e.g., competing reactions in a fluid should not be confused with operations in a vacuum). Molecular manufacturing concepts often deal with sufficiency; i.e., the design is sufficient—not perfect—but this is not a fundamental flaw.

checkpoints and development risks should be considered for planning purposes. An incremental support plan should not at this point be tied into any particular concept of assembler, nor into one path. Reliance on the SFM is only one of several methods for fabricating products from the bottom-up, and a spectacular chemical-reaction-building sequence with SFMs would be a milestone, but parallel checkpoints (or milestones if the pathways become firmer in the next few years) for assembly might be found with several chemical synthesis avenues or biological analogues that can be considered for nanofabrication.

Progress in miniaturizing probe arms, using probes to not only position molecules but to force a sequence of distinct chemical reactions, learning the process of macromolecular folding and how to self-assemble polymers and films, and progress in molecular modeling are all important topics to address and quantify practical limiting factors.[6] It may be possible to achieve sufficient results should only one of the routes prove viable, so a multipath approach to molecular manufacturing seems advisable for the near term. The use of arrays of probes is linked to top-down nanotechnology based on familiar semiconductors; therefore, versatility in probe design should probably be encouraged. (In addition to STM and AFM variants, exploration of utilizing other forms of SFM should be encouraged.)

Plans for developing manufacturing tools, and ultimately products, will need to incorporate a process for examination of the resulting products in a number of steps. A termination point for molecular nanotechnology-based manufacturing could result from any of several steps failing. All intermediate steps, whether sequential or combined, must be feasible for an engineering route to be plausible. The following are steps that should be demonstrated to produce useful systems or objects:

- Produce material parts at the nanoscale.

- Process material parts into components at the nanoscale.

- Order molecular components into a structure and interconnect.

- Interface system components with the macroenvironment.

---

[6]One of the major research efforts needed is to understand the details of the interaction of (fundamentally electromagnetic) forces that can hold molecules to a surface or a probe tip, and learning not only how to bias a tip with respect to a surface to manipulate particles (as is already demonstrated in many examples of "direct writing" with atoms or globs of atoms) but also how to orient and retain molecules for sequences of chemical reactions to occur.

- Control a massive collection of miniature parts and systems.
- Provide a power system.

The first two bullets may be separate steps or may happen simultaneously under some fabrication paths—e.g., molecular self-organization. The distinction between production of nanomaterials and nanomachining of the materials is not always present.

A full systems integration is a critical concern. The biological cell, for instance, is enormously complex, organized by a system of macromolecular structures. The interconnection, interface, and control steps are major concerns with systems depending on an enormous collection of nanoscale components or features. Error mechanisms and rates must be understood and controlled or compensated for.

Any nanomachine must have a power system—an energy source, storage mechanism(s), transmission system(s), thermal management, and handling (control, packaging, and power conversion). This is a point that has sometimes been glossed over in discussions of developing practical molecular nanosystems. The energy source may be a macroscopic generator if the nanosystem is equipped for reception and conversion; e.g., photosynthesis or ultrasound-driven reactions. Intermediate storage of energy for access might be necessary. Nanobattery chemistry is an area that could be considered for development. Thermal management is very important—in ordinary microelectronics, the cooling of chips as capability increases is a major design consideration. Transmission of power through the system depends on the design concept—in the case of photosynthesis, intermediate donor and receptor molecules serve to relay energy supplied by the initial photon. Some potential mechanisms for nanodevices, such as bioenergetics, may allow for efficient, "adiabatic" processes in subsystems. In addition to considering the role of transfer of electrons or photon effects in molecular devices, osmotic bioenergetic mechanisms that involve the flow of ions across membranes and chemical storage or intermediaries can be considered as candidates for mimicry (perhaps in conjunction with interfaced molecular "wires" for electrons). The regeneration of the adenosine tri phosphate (ATP) is very important to the functions of nutrition and respiration, for example, in living organisms.

Alarms have been raised over the potential horror of success in molecular nanotechnology. There is concern that molecular-based nanosystems could lead to problems with human health, the environment, crime, or military use by powers hostile to U.S. interests.[7] These are serious topics that would require study if advanced molecular nanotechnology applications emerge.

---

[7]Examples include the development of lethal and specific poisons, potent new drugs that might usher in unexpected modes of drug abuse, uncontrollable viruses wreaking havoc on health, sabotage of utilities or facilities penetrated by hordes of miniature robotic attackers, or destitution of the environment by a destructive, replicating miniature machine out of control. The worst fears about the potential of molecular nanotechnology consider the introduction of potential new doomsday scenarios ranging from competition with a superior artifical intelligence to the total physical destruction of ourselves and possibly the biosphere—sometimes called the "grey goo" scenario. Concerns about destructive consequences, as is the case with hope for constructive purposes, must ultimately weigh engineering reality against fantasy.

4. National and International Research Efforts

Categorization

Three methods were used to acquire information on laboratories conducting nanotechnology research relevant to molecular manufacturing. First, a literature search was conducted with on-line databases, and information acquired from the articles and books found therein. Second, a search was done on the National Science Foundation's (NSF) on-line database of projects to gather information on nanoscale research and development grants issued across NSF's many divisions and departments. Finally, an internet posting invited researchers in the field to help identify centers of research. Each center was then categorized by sector (government, industry, or academia), and by the research focus they were found to be pursuing (nanomeasurement, nanomachining, and/or molecular materials properties). More than 200 centers worldwide were identified (a list of these centers and their countries of origin is provided in the appendix).

The number of centers identified for a given country broken down by sector performing research, or alternatively by research focus, would not always coincide since a given center can conduct research in more than one application area. It is sometimes difficult to distinguish between centers focused primarily on top-down research and those studying problems more specifically of interest in molecular nanotechnology-based manufacturing.

A second limitation on the survey is that there are likely to be many centers not represented in this type of survey that are nevertheless active in research contributing to the science of molecular manufacturing. This search is presumed to have missed a number of important activities in other nations as well as some relevant work in the United States, particularly in chemistry and biology departments, despite the internet posting. The prospect of omissions is an error on the conservative side of estimating widespread activity relevant to a base for molecular nanotechnology.

Figure 5 provides a tally of the research centers.[1] In terms of the number involved in nanotechnology relevant to molecular manufacturing, the United States appears to be a leader having approximately half of all the identified centers.

However, more than two-thirds of these U.S. centers are in academia and often consist of a few professors in one or two departments. The low numbers for the rest of the countries may not be indicative so much of a lack of support in nanoscale research as an indication of our search limits. The key point is that a

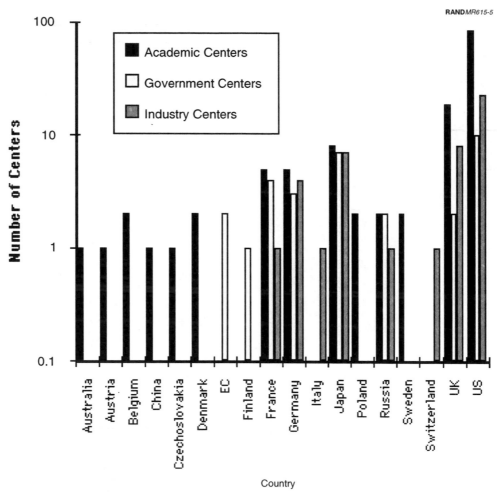

NOTE: EC is the European Commission.

Figure 5—Research Centers in Nanotechnology by Sector and Country

---

[1]A logarithmic scale has been chosen for ease of representing many nations with widely different numbers of identified centers. Thus, small differences in the upper portion of the chart are more significant than may be apparent.

number of countries are engaged in some level of activity. Most of the countries with few explicitly identified centers (i.e., Australia, Austria, Belgium, China, etc.) have one or two centers in academia.

The United States and the United Kingdom have fairly similar distributions by sector with a heavy emphasis on academia. Japan's centers are remarkably equally distributed among sectors, reflecting the close cooperation traditionally found in Japanese technology efforts (however, we have not unfolded the numerous individual efforts in Japan—see below).

Figure 6 demonstrates the presence of research emphasis on both nanomachining and molecular materials methods (i.e., as defined in Figure 2), with varying ratios by nation. In terms of applications, U.S. centers seem to reflect a strongly diverse base for relevant molecular methods (unbridled inclusion of biotechnology efforts would easily inflate the estimate). In the United Kingdom, the centers seem equally interested in nanomachining and projects oriented toward molecular methods and nanomaterials properties. Japanese centers show a strong representation in nanomachining, but as we shall presently indicate, Japan is stronger in molecular materials than the head count implies. As the rest of the world develops expertise in biotechnology or materials engineering for industry, it can be expected that the molecular methods criteria will show the gap between identified leaders and others beginning to close, especially as instrumentation for high-precision examination becomes more available.

Leading Nations

Beyond a simple numerical analysis of the number and categorizations of the centers discovered, there is a further categorization into those nations with national centers for nanoscale research and those nations with coordinated national programs in nanoscale research and molecular manufacturing. China, Denmark, France, Germany, Japan, Russia, Sweden, the United Kingdom, and the United States have what can be termed "national centers of excellence" in this field. China has the Laboratory of Molecular and Biomolecular Electronics at Southeast University. Denmark has the Centre for Interdisciplinary Studies of Molecular Interactions (CISMI) at the University of Copenhagen. France has both the University of Strasbourg and the Laboratoire de Chimie Supramoleculaire. Germany has the Max-Planck Instituts fur Polymerforschung and fur Festkorperforschung and the Fraunhofer Institute for Solid State Technology. Japan has the Tsukuba Nanotechnology Research Facility and Electrotechnical Laboratory, including the Super Molecular Science Division, the Material Science Division, and the Frontier Technologies Division; and the

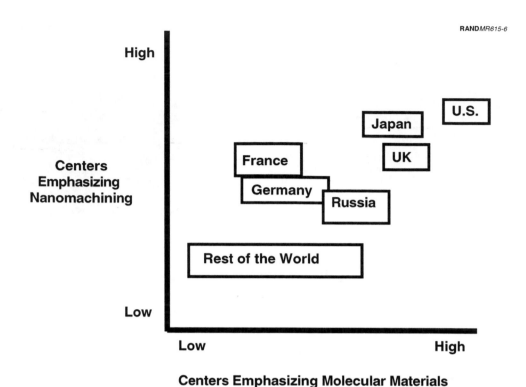

RAND*MR615-6*

Figure 6—Comparative Research by Area Pursued

Okazaki National Research Institutes, including the Institute for Molecular Science and the Department of Applied Molecular Science. Russia has the Institute on Nanotechnology and Nanoelectronics. Sweden has the Lund University Nanometer Structure Consortium. The United Kingdom has the University of Warwick Center for Nanotechnology and Microengineering and the National Physical Laboratory (NPL). Finally, the United States has the Cornell University National Nanofabrication Facility, numerous NSF Science and Technology Centers (STCs), NSF Engineering Research Centers (ERCs), and NSF Cooperative Research Centers (CRCs) with work related to the field.

Though many nations have developed centers in nanotechnology and molecular manufacturing, there are few coordinated national programs in the field. Japan has a long-standing national program through their Science and Technology Agency (STA) known as ERATO, and two more recent projects initiated through the Ministry of Trade and Industry (MITI). STA's ERATO project has been under way since 1981, while MITI's Angstrom Technology Project (ATP) and Quantum Functional Devices Project (QFD) have been initiated only in the past several years.

Besides these Japanese national projects, the United Kingdom has had the Link Nanotechnology Project under way since the late 1980s; and the European Commission (EC) has recently started the NEXUS Nanotechnology project.

The United States and Japan: A Comparison of Support

Data on national budgets for relevant research and development was derived from several major sources. Our principal estimates on U.S. spending come from the summation of NSF grants for nanoscale research. Data on Japanese spending comes from the announced budget of MITI projects and the announced average value of STA projects multiplied by the number of potentially relevant STA projects. The STA ERATO project has consisted of 33 research projects from 1981 through 1997. Eighteen of these have been completed as of 1993. While many of these projects are explicitly devoted to efforts related to nanotechnology and molecular manufacturing, many others are of only ancillary applicability or are completely not applicable to the field. If we directly compare the sum of the NSF molecular manufacturing-relevant spending to Japan's total MITI and ERATO spending, we find that the United States through the NSF (the only agency with data that was accessible for the search) has committed $150,714,931 (then-year dollars) from 1985 through 1997 on nanotechnology ($11,593,456 per year) and Japan has committed substantially more.[2]

Several ERATO projects have already been completed—these projects tend to be highly exploratory, with definite end-dates attached. Although many of these are not directly applicable in molecular manufacturing development efforts, the research results are often relevant.

Examples of relevant ERATO projects that have been undertaken are the following:

- Nano-mechanism (1985–1990, Yoshida. The development of a laser interferometer system for controlling positioning through rolling ball guide to the nanometer level).

- Molecular dynamic assembly (1986–1991, Hotani).

- Molecular architecture (1987–1992, Kunitake).

- Atomcraft (1989–1994, Aono. STMs used to position atoms on surfaces).

---

[2]There are a substantial number of U.S. efforts supported by additional agencies—such as the Advanced Research Projects Agency, the Department of Energy, the National Institutes of Health, etc.—that support a base for molecular nanotechnology and should be included in any detailed assessment of the relative standing of nations. Our intention is not to provide a precise ratio of national efforts, but rather to estimate if parity exists.

- Protein array (1990–1995, Nagayama. Investigation of the use of proteins as building blocks).

- Molecular catalysis (1991–1996, Yoshimura. Catalysis is a method of overcoming the activation energy required for chemical/biological reactions and is relevant to developing basic scientific understanding helpful in some molecular manufacturing).

- Millibioflight (1992–1997, Kawachi. Bioenergetics, biodynamics and control, and fluid dynamics of tiny organisms may in principle yield results that could be combined with nanotechnological devices).

The MITI funding requires a 30 percent additional industry commitment. Thus, the $245,000,000 that MITI has devoted to the field of nanotechnology, with a very strong molecular nanotechnology orientation, has fostered at least an additional $73,500,000 investment on the part of Japanese industry. This investment is added onto the investment by Japan's Science and Technology Agency. If we estimate that only one-third of that effort is related to molecular manufacturing, we find a total STA budget from 1981–1997 of $220,000,000. This leads to a total Japanese investment of $538,500,000.

If we assume that the rest of the U.S. government's investment is equal to the NSF investment, and that industrial investment is equal to NSF's investment (giving more credit to industry and government than may be due), we find a total investment in the United States over this period of $452,144,739 from 1985–1997. This is comparable to the Japanese effort and outdistances other nations.

Though these numbers are highly speculative, they do represent a general finding. For example, by summing grants to the Cornell University National Nanofabrication Facility that may be relevant to molecular manufacturing, we identify over the period of 1987–1993 that the center received $11,895,955. Thus, for each of those seven years, this effectively averaged $1,699,422. However, the average Japanese STA project lasts five years and is budgeted at approximately $20,000,000 ($4,000,000 per year).

Using this approach we find that the U.S. National Nanotechnology Facility is budgeted at only 43 percent of the average Japanese project. However, the 1994 Research Centers Directory indicates that the U.S. National Nanofabrication Facility is budgeted at $4.8 million (Cichonski, 1994). This would make the U.S. National Facilities funding similar to the average Japanese project, although the relevance of the U.S. focus for molecular-nanotechnology may be different. We observe that although the United States has a great many centers in nanotechnology and molecular manufacturing, Japan has larger centers that are funded at a significantly higher rate than their U.S. counterparts and are coordinated by a dedicated national effort.

5. Conclusions and Recommendation

Competitive Status

Though it is an immature field, molecular manufacturing has a strong and growing scientific and technical reality that will likely lead to significant scientific and economic benefits over the next few decades. Though the United States currently has a competitive advantage in this field, U.S. research and development efforts are decentralized and uncoordinated, and individual efforts are often at levels lower than U.S. competitors' centers. Increased funding and coordination could enhance the probability that the United States will remain competitive in this field and realize the potential benefits to be reaped from molecular manufacturing. However, significant near-term research steps need to be demonstrated. Our analysis of international centers of excellence has led to several policy-relevant conclusions.

The United States is currently a leader in the various approaches to nanotechnology. The U.S. effort may be overly centered in academic laboratories, which will have to advance to a manufacturing orientation at some point if applications of molecular nanotechnology do show rapid expansion. This effort is strong in the basic studies of materials properties. Extensive organization or coordination of molecular nanotechnology at the basic and applied research level is lacking in comparison to Japan's programs.

For molecular-based nanotechnology specifically, the sum of the U.S. efforts is on a par with efforts in Japan and probably ahead of those in Europe. This leading position is unstable and could be challenged in the near future if development breakthroughs appear. It is for this reason that we are led to suggest a cautious but advisable step to take at this time is a very comprehensive assessment of both the state of the art of molecular fabrication methods and the potential being realized in basic nanotechnology research.

Fundamental Research Support

It is widely recognized that an enormous number of research areas might make contributions to molecular nanotechnology. There are several key areas of science and technology to watch for developments. Promising efforts in these could be supported as building blocks to molecular manufacturing. Research areas include (but are certainly not limited to)

- macromolecular design and folding
- self-assembly methods
- catalysis (inorganic, enzyme, and other)
- dendrimers, fullerenes, and other novel chemical structures
- bioenergetics, nanobatteries, ultrasound-driven chemistry
- semiconductor-organic/biological interfaces
- miniaturization and massive parallelism of SFM
- molecular modeling tools.

Should molecular-nanotechnology-based advanced manufacturing fail to deliver the ambitious goals being proposed, research in these areas will nevertheless provide many direct benefits. SFM tools have many uses such as atomically resolved imaging of biological and other materials; the study of frictional forces with AFM; use of massively parallel nanotips in flat panel displays, high-density information storage, and other nanoelectronics; or for improving components used in information systems, the application of electrostatic force microscopes to probe chip dopants, or magnetic force microscopes to probe magnetic media. The design and synthesis of macromolecular and film structures can be useful in many applications, such as artificial photosynthetic devices for energy conversion or material production, the tailoring and improvement of therapeutics, smart structures for improving aircraft aerodynamic performance or to report and respond to failures in aircraft or automobile systems. A large fraction of industrial processes presently depend on catalysis, and exploration of novel chemical structures such as caged molecules (e.g., zeolites) offers potential for breakthroughs in many areas (e.g., lubrication, superconductivity, and environmental treatment).

Although the historic focus of nanotechnology has been on nano*meter* measurement and manipulation, miniature phenomena encompass more than just spatial dimensions. In conjunction with learning how to operate and manufacture with ultrasmall phenomena, the application of ultra*fast* phenomena (i.e., temporal events at the pico- or even femtosecond scale) in atoms or molecules is an important research area. This might be of direct use in novel molecular computer architectures, or as an aid in examining biological or chemical interactions and manipulating molecules for manufacturing goals. Research areas to examine include cytoskeletal structures within the cells of organisms, control of chemistry with picosecond or shorter laser pulses, and the various means of compactly generating short, high-intensity beams.

A comprehensive review of nanotechnology for advanced molecular manufacturing would not directly bring the realization of tentative concepts to fruition but would organize detailed approaches and identify more concrete potential cutoff points.

Incremental Actions Needed

Because there are alternative paths to an advanced molecular manufacturing capability, rather than focusing on one extremely ambitious goal, a number of steps should be considered. Milestones to identify include not only the proof of principle of a highly capable assembler (or replicator), but also the many interim products that are valuable in themselves and are expected to continue to arise from molecular nanotechnology developments just as microelectromechanical systems and microelectronics before them produced an incremental explosion of capability.

We recommend evaluation of various means to promote coordination at the basic and applied research level beyond the present situation. Given the nascent status of molecular manufacturing, it is unclear as yet which direction in molecular manufacturing users will best prosper from, and options should be left open at the basic and applied research level for various approaches to molecular manufacturing. It might be expected that if sufficient research breakthroughs as we have described are eventually achieved, then manufacturers would be motivated to pursue development of applications.

The following options impart a number of diverse, often conflicting, alternatives that might be considered to address research oversight in molecular manufacturing and nanotechnology. Policymakers may opt for one of various approaches, from benign noninterference to strictly coordinated programs. Analysis was not attempted to rank the program options by any rationale, but marginal comments are presented.

Delay or Forego New Policy Action

Option 1: Maintain Laissez-Faire Policy

One option open to policymakers is to maintain a laissez-faire environment toward molecular manufacturing in both the public and private sectors. This policy would maintain *de facto* control of molecular manufacturing prioritization in the decentralized peer-review process in use at the NSF, the National Institutes of Health (NIH), etc. Thus, this policy is likely to maintain the current focus in academia and runs a risk of keeping results "in the lab" with relatively little involvement by, and dissemination to, the private sector.

Option 2: Technology Assessment and Feasibility Analysis

A step beyond option 1 would be to carefully investigate the need for new actions, dedicating resources to molecular nanotechnology. Decisionmaking on funding levels and priorities could be greatly improved by having an independent agent (e.g., the American Physical Society or the National Academy of Sciences) perform a **comprehensive** assessment of the state of current nanotechnology tools and molecular manufacturing capabilities, funding levels, and organization. The resulting assessment could provide an objective and detailed estimate of the potential benefits of pursuing molecular manufacturing, cost of pursuing molecular nanotechnology-based manufacturing, and likely benefits to be realized in the near term. This option can be recommended as a hedge against "technological surprise."

Establish Coordinator(s)

Option 1: NSF Coordinator for Nanoscale Research and Development (R&D)

Current NSF nanoscale R&D are performed across the entire range of NSF divisions, directorates, and departments. This lack of coordination can lead to redundancy and a lack of information sharing. Appointing an NSF coordinator for nanoscale R&D to review grant funding and program results could go a long way to increase the efficiency of R&D in molecular manufacturing and increase the visibility of (molecular and other) nanotechnology-based manufacturing as a viable field of R&D.

Option 2: National Institute of Standards and Technology (NIST) Coordinator for Nanotechnology and Molecular Manufacturing

NIST molecular manufacturing research is currently undertaken in at least three divisions. As explained in option 1, appointing a NIST coordinator could enhance the efficiency of NIST funding for molecular manufacturing research and further increase the visibility of molecular nanotechnology-based manufacturing as a viable field of R&D.

Option 3: NIH Coordinator for Protein Engineering and Molecular Biotechnology

NIH has numerous molecular biotechnology and protein engineering projects. As above, coordinating these efforts could work to help maximize the efficiency of federal spending.

Option 4: Advanced Research Projects Agency/Department of Defense (ARPA/DoD) Coordinator for Molecular Manufacturing for Civilian and Defense Applications

An ARPA coordinator could ensure that ARPA projects relevant to nanotechnology and molecular manufacturing (such as its work on advanced semiconductor technology) receive appropriate funding and are appropriately prioritized and coordinated.

Option 5: National Aeronautics and Space Administration (NASA) Coordinator for Molecular Manufacturing for Aerospace Applications

A NASA coordinator could ensure that the agency is appropriately pursuing molecular manufacturing technologies of benefit to the aerospace community, and that such efforts are efficiently undertaken.

Option 6: Department of Energy (DoE) Coordinator for Nanotechnology and Molecular Manufacturing

DoE laboratories have significant experience in nanomeasurement, nanomachining, and nanomaterials processing. A coordinator could ensure that their efforts are not redundant, are properly funded and prioritized, and are effectively disseminated to the private sector.

Option 7: Coordinating Committee on Nanoscale R&D and Molecular Manufacturing

A cross-agency coordinating committee could further maximize the efficiency of federal R&D in molecular manufacturing by reducing redundant efforts across agencies, by ensuring appropriate prioritization, and by increasing information flows between research communities.

Establish New Program(s)

Option 1: U.S. Molecular Manufacturing Research Program

A U.S. Molecular Manufacturing Research Program, akin to the U.S. Global Change Research Program or the High Performance Computing Initiative, might more effectively maximize the efficiency of federal R&D by requiring joint planning and prioritization across agencies involved in molecular manufacturing research. Such a program would also enhance private sector R&D by increasing the visibility of molecular manufacturing and the perceived viability. However, such a top-down and structured approach runs the greatest risk of reducing the benefits resulting from competing researchers and institutions.

Option 2(a): Enhance NSF Funding for Nanoscale Research and Development

One way to forestall a "funding gap" between the United States and the "rest of the world" would be to increase NSF funding dedicated to nanoscale R&D. However, such increased funding would likely require the appointment of an NSF coordinator to ensure that the earmarked funding was efficiently and appropriately spent given the decentralized nature of NSF molecular manufacturing efforts (note that this finding holds true for increasing funding at the other agencies).

Option 2(b): NSF Program in International Nanoscale R&D

An NSF program in international nanoscale R&D could improve U.S. competitiveness by increasing the awareness of U.S. researchers about the findings and capabilities of other nations.

Option 3: NIST Program in Nanotechnology and Molecular Manufacturing

Increasing funding for molecular manufacturing at NIST could improve U.S. competitiveness given NIST's commercial orientation.

Option 4: NIH Interdisciplinary Program in Biotechnology and Nanotechnology

The United States would appear to have a significant advantage in protein engineering and molecular biotechnology over foreign competitors. Enhanced funding for such research could ensure maintenance and even improvement of this competitive advantage.

Option 5: ARPA/DoD Program in Molecular Manufacturing for Civilian and Defense Applications

Enhanced funding at ARPA for molecular manufacturing research could lead to significant advances given ARPA's history with computer innovations and semiconductor manufacturing and could ensure appropriate consideration and development of nanotechnology with defense-related applications.

Option 6: NASA Program in Molecular Manufacturing for Aerospace Applications

Enhanced NASA funding for molecular manufacturing research could lead to advances in aerospace materials and ensure that potential aerospace applications of molecular manufacturing were appropriately explored.

Option 7: DoE Federal Laboratory Program in Nanotechnology and Molecular Manufacturing

The DoE National Laboratories have a long history in research relevant to molecular manufacturing and provide a natural focus for enhanced funding to forestall a capability gap with foreign competitors. Such funding could lead to significant advances in both top-down and bottom-up molecular manufacturing.

Option 8: Small Business Innovative Research (SBIR) Program in Molecular Manufacturing

To ensure that innovative smaller firms are involved in R&D activities related to molecular manufacturing, a focused SBIR program can be created to provide access to necessary funding and facilities. Such a program could increase the activity of industry in the field and ensure commercialization of potential products. However, some promising commercial avenues may not be explored with such a program given the low funding level of SBIR grants and the limited capabilities of small businesses.

Cooperative Efforts

Option 1: National Consortium for Molecular Manufacturing

The creation of a national consortium for molecular manufacturing could provide for enhanced involvement of large firms; significant increases in information sharing and cooperative research between large and small firms, academia, and government laboratories; enhanced funding for molecular manufacturing research from both the private and public sectors; and increased research efficiency by reducing redundant efforts across all aspects of the research community.

Option 2: International Cooperation

The creation of an international research center might follow in the image of the European CERN (laboratory for particle physics) or the U.S. Fermilab high-energy facilities. There are already a number of U.S. research scientists who participate in Japanese nanotechnology research projects.

Recommendation

Given the combination of potentially large far-term payoffs with major current technical uncertainties, a vital first step would be a *detailed* and *objective* technology assessment examining the current status and likely prospects of molecular technology. The working group that performs this assessment should consist primarily of biotechnology experts, chemists, computer scientists, electrical engineers, materials scientists, mechanical engineers, and physicists to represent the contributing disciplines. The challenge is to bring together leading experts who can participate in unbiased but informed analysis of a multidisciplinary topic.

Pending a positive finding, a number of options give alternative means of providing needed organization depending on the most credible goals identified by a major technology assessment. The role of government in developing applications-oriented molecular manufacturing technologies can be examined within a framework to establish policies maximizing free market incentives for such developments. A negative finding from an objective technology assessment would strongly indicate that the current levels of funding and structures for basic scientific research in molecular-based nanotechnology are appropriate.

Appendix

Research Centers in Nanotechnology and Related Areas by Nation

The following is a limited representation of centers that have performed research relevant to nanotechnology in general. It is expected that this reflects national strength for participating in future precision molecular system developments. Clearly, the ability to identify U.S. interests outstrips our identification of foreign centers and correspondingly skews conclusions—our database searches and internet posting would not have had either universal reach nor a flat sampling probability by nation. In particular, efforts outside of the United States, Japan, and Europe undoubtedly exceed our identification results. However, this initial effort may be considered broadly representative of distinctions among nations that show their level of interest and support for nanotechnology, with the additional but very important caveat that distinction is not made between small research efforts and entries that represent large investments. Therefore, a simple numerical summation does not show the entire picture. To be somewhat more instructive, limited organizational information is included—in some cases, multiple efforts are to be found at one location.

Country	Organization
Australia	Royal Melbourne Institute of Technology
Austria	Vienna Technical University, Institute Allgemeine Physics
Belgium	Service de Chimie des Materiaux Nouveaux
Belgium	Universite de Mons-Hainaut, Departement des Materiaux et Procedes
China	Southeast University, Laboratory of Molecular and Biomolecular Electronics
Czechoslovakia	Slovak Technical University, Department of Inorganic Chemistry/Microelectronics
Denmark	Technical University of Denmark, Chemistry Dept.
Denmark	University of Copenhagen, The Centre for Interdisciplinary Studies of Molecular Interactions (CISMI)
EC	European Molecular Biology Laboratory
EC	NEXUS Project
Finland	VTT (Technical Research Centre of Finland), Machine Automation Laboratory
France	CEN-Saclay, Service Chimie Moleculaire

France	Centre d'Etudes Nucleaires de Grenoble, Departement de Recherche Fondamentale sur la Matiere Condense, Laboratoire de Chimie de Coordination
France	CNES
France	CNRS-ULP, Institut Charles Sadron
France	ESPCI-CNRS
France	Institut de Chimie Organique
France	Institut de Physique et Chimie des Materiaux de Strasbourg
France	Laboratoire d'Electronique Phillips
France	Universite Bordeaux
France	University of Strasbourg, Laboratoire de Chimie Supramoleculaire
Germany	Eppendorf-Netheler-Hinz
Germany	Fraunhofer Institute for Solid State Technology
Germany	IBM Munich
Germany	Joh. Gutenberg Universitat, Institut fur physikalische Chemie
Germany	Max-Planck Institut fur Polymerforschung and fur Festkorperforschung
Germany	Messerschmitt-Boelkow-Blohm GmbH
Germany	Scientific Electronics Munchen GmbH
Germany	Techische Universitat
Germany	TH Darmstadt, Institute Hochfrequenztech
Germany	Universitat Berlin
Germany	Universitat Stuttgart, Physikalisches Institut and Institut fur Organische Chemie
Germany	Universitat Tubingen, Institut fur Physikalische und Theoretische Chemie
Italy	Bracco Industria Chimica SpA
Japan	Ashikaga Institute of Technology
Japan	Hitachi Central Research Laboratory
Japan	Japan Electronics Optics Laboratory
Japan	Kyoto University, Dept. of Hydrocarbon Chemistry
Japan	Matsushita Corp.
Japan	MITI, Angstrom Technology Project
Japan	MITI, Quantum Functional Devices Project
Japan	NEC, Inc. (Advanced Device Research Lab, Functional Devices Research Lab)
Japan	Nippon Telephone and Telegraph
Japan	Okazaki National Research Institutes, Institute for Molecular Science, Department of Applied Molecular Science
Japan	OMRON Corporation, Manufacturing Technology R&D Lab
Japan	Optoelectronics Technology Research Laboratory
Japan	Osaka University
Japan	Protein Engineering Research Institute
Japan	Saitama University
Japan	Science University of Tokyo

Japan	Seiko-Epson Corporation
Japan	Shizuoka University, Dept. of Precision Engineering
Japan	STA, JRDC, ERATO (Exploratory Research for Advanced Technology), numerous projects
Japan	Tokyo Institute of Technology, Department of Biomolecular Engineering and Department of Mechanical Engineering for Production
Japan	Toshiba Corporation
Japan	Tsukuba Nanotechnology Research Facility and Electrotechnical Laboratory, Super Molecular Science Division, the Material Science Division and the Frontier Technologies Division
Netherlands	University of Leyden
Netherlands	University of Nijmegen, Faculty of Science, Computer Assisted Organic Synthesis and Computer Assisted Molecular Modeling Center
Poland	Stanislaw Staszic University of Mining and Metrology, Dept. of Carbon Energy Chemistry and Sorbent Physical Chemistry
Poland	Technical University of Wroclaw, Institute of Organic and Physical Chemistry
Russia	Institute on Nanotechnology and Nanoelectronics
Russia	Molecular Device and Technology
Russia	MV Lomonsov State University, Center for Advanced Technologies
Russia	Russian Academy of Science, Institute of Electrophysics
Russia	USSR Academy of Sciences, Institute of Crystallography
Sweden	Linkoping University, Department of Physics
Sweden	Lund University, Department of Solid State Physics, Nanometer Structure Consortium
Switzerland	IBM, Zurich Research Lab
UK	Biodigm
UK	Birmingham University
UK	City University, Northampton Square, Measurement and Instrumentation Centre
UK	Cranfield Institute of Technology, Centre for Molecular Electronics
UK	GEC Marconi Material Materials Technology Ltd.
UK	Glasgow University
UK	Griffith University, Division of Science and Technology
UK	ICI Wilton Materials Research Centre
UK	Link Nanotechnology Programme
UK	Marathon Oil and Dow Chemicals
UK	Marconi Research Centre
UK	Merck, Ltd., Organic Development Department
UK	National Physical Laboratory (NPL), Div. of Mechanical and Optical Metrology
UK	Oxford Applied Research, Ltd.
UK	Oxford Molecular
UK	Queen Mary and Westfield College, Physics Department

UK	Queen's University
UK	UMIST, Dept. of Instrumentation and Analytical Science
UK	University of Birmingham, School of Manufacturing and Mechanical Engineering and School of Metallurgy and Materials
UK	University of Cambridge, Dept. of Material Science and Metallurgy
UK	University of Liverpool, Dept. of Material Science and Engineering
UK	University of London Imperial College of Science, Technology and Medicine, Dept. of Electrical & Electronic Engineering, Optical and Semiconductor Device Section
UK	University of Loughborough
UK	University of Nottingham, Dept. of Pharmaceutical Science, Biophysics and Surface Analytics Lab
UK	University of Oxford
UK	University of Salford
UK	University of Wales, School of Electrical Engineering and Computer Systems
UK	University of Warwick, Dept. of Engineering, Center for Nanotechnology and Microengineering
US	3M
US	Aerospace Corp.
US	American Red Cross, Jerome Holland Lab
US	American Vacuum Society, Chemistry Division
US	Appalachian State University, Dept. of Physics and Astronomy
US	Apple Computer
US	Argonne National Laboratory
US	Arizona State University
US	AT&T Bell Labs - Murray Hill
US	Autodesk
US	Batelle Northwest Laboratories, Chemical Sciences
US	Boston University
US	Brown University, Division of Engineering, Cooperative Research Center for Thin Film and Interface Research
US	California Institute of Technology, Center for the Development of an Integrated Protein and Nucleic Acid Biotechnology and Center for Molecular Biotechnology
US	California Institute of Technology, Dept. of Physics and Dept. of Chemical Engineering, Materials and Molecular Simulation Center
US	Carnegie Mellon University, NSF STC, Center for Light Microscope Imaging and Biotechnology and Data Storage Systems Engineering Research Center
US	Case Western Reserve, Cooperative Research Center for Molecular and Microstructure of Composites
US	Clark Atlanta University, Research and Resource Center for Electron Microscopy
US	Clarkson University, Dept. of Chemical Engineering
US	Columbia University

US	Cornell University, National Nanofabrication Facility
US	CUNY City College, Center of Analysis of Structures and Interfaces (CASI)
US	Dartmouth College
US	Dow Chemical Corporation
US	DuPont
US	Eastern Michigan University, Cooperative Research Center for Coatings
US	Eli Lilly Corporation
US	Exxon Corp.
US	GA Technology Research Corporation (GTRI)
US	Georgia Institute of Technology, Center for Computational Materials Science
US	Harvard University
US	Howard University, Materials Science Research Center of Excellence
US	IBM Almaden Research Center
US	Institute for Molecular Manufacturing
US	Iowa State University
US	Johns Hopkins University, Applied Physics Lab and Center for Biophysical Studies on Macromolecular Assemblies
US	Kansas State University
US	Kent State University, NSF STC, Center for Advanced Liquid Crystalline Optical Materials
US	Lehigh University, Dept. of Chemical Engineering, Polymer Interface Center
US	Los Alamos National Lab
US	Massachusetts Institute of Technology, Dept. of Materials Science and Engineering, Dept. of Electrical Engineering and Computer Science, Dept. of Physics, and Biotechnology Process Engineering Research Center
US	Materials Research Group
US	Materials Research Society, Materials Science and Metals Science and Engineering Division
US	Meharry Medical College, Center for the Study of Cellular and Molecular Biology
US	MMI
US	Molecular Biosystems Inc.
US	Montana State University, Engineering Research Center: Interfacial Microbial Process Engineering
US	Nanotronics
US	NASA Ames Research Center
US	National Institute of Diabetes and Digestive and Kidney Diseases
US	Naval Research Laboratory
US	New Mexico Institute of Mining and Technology
US	NIST, Electron Physics Div., Factory Automation Systems Div., and Precision Engineering Div.

US	North Carolina State University, Department of Materials Science and Engineering, Advanced Electronic Materials Processing Engineering Research Center and Precision Engineering Center
US	Northwestern University, Department of Chemistry and Materials Science and Engineering, Materials Research Center
US	Novasensor
US	Oklahoma State University
US	Optical Society of America
US	OTA
US	Park Scientific Instruments
US	Particle Technology, Inc.
US	Pennsylvania State University at University Park, Department of Ceramic Science and Engineering, Materials Characterization Laboratory and Cooperative Research Center for Dielectrics
US	Princeton University
US	PSI Technology Company
US	Purdue University
US	Rensselaer Polytechnic Institute, Dept. of Physics
US	Rutgers University at New Brunswick, Cooperative Research Center for Ceramics Research
US	SAIC
US	Sandia National Laboratory, Microelectronics and Photonics Directorate
US	Scripps Research Institute
US	Stanford University, Dept. of Materials Science and Engineering and Edward L. Ginzton Lab
US	Stavely Sensors
US	SUNY at Buffalo, Dept. of Electrical and Computer Engineering, Center for Electrical and Electro-optical Materials
US	SUNY at Stonybrook, Dept. of Materials Science and Engineering
US	Syracuse University, Dept. of Chemistry, Science and Technology Research Center for Molecular Electronics
US	Texas A&M Univeristy
US	Tufts University
US	United Engineering Trustees, Inc., Manufacturing Research
US	University of Arizona, Dept. of Materials Science and Engineering and Optical Sciences Center
US	University of California at Berkeley, Division of Neurobiology and Lawrence Berkeley Laboratory
US	University of California at Davis, Dept. of Mechanical and Aeronautical Engineering and NSF STC, Center for Engineering Plants for Resistance Against Pathogens
US	University of California at Irvine
US	University of California at San Diego, Cooperative Research Center for Ultra-High Speed Integrated Circuits and Systems
US	University of California at Santa Barbara, NSF STC, Center for Quantized Electronic Structures

US	University of Cincinnati, Dept. of Materials Science and Engineering
US	University of Colorado, Opto-electronic Computing Systems Engineering Research Center and Cooperative Research Center for Separations Using Thin Films
US	University of Connecticut, Cooperative Research Center for Grinding
US	University of Florida
US	University of Illinois at Urbana-Champaign, Dept. of Materials Science and Engineering, NSF STC, Center for Magnetic Resonance Technology for Basic Biological Research and Compund Semiconductor Microelectronics Engineering Research Center and Beckman Institute
US	University of Iowa
US	University of Kentucky, Dept. of Chemistry
US	University of Maine, Laboratory for Surface Science and Technology
US	University of Massachusetts at Amherst, Microscopy and Imaging Center
US	University of Michigan at Ann Arbor, Dept. of Chemistry and NSF STC, Center for Ultrafast Optical Science
US	University of Minnesota, School of Medicine and Interfacial Engineering Research Center
US	University of Nebraska at Lincoln
US	University of New Mexico, Dept. of Chemistry and Cooperative Research Center for Micro-Engineered Ceramics
US	University of North Carolina at Chapel Hill
US	University of North Carolina at Charlotte, Department of Mechanical Engineering
US	University of North Texas, Dept. of Physics and Cooperative Research Center for Nanostructural Materials Research
US	University of Oregon at Eugene
US	University of Rhode Island
US	University of Rochester, NSF STC, Center for Photoinduced Charge Transfer
US	University of South Carolina at Columbia, James F. Byrnes International Center
US	University of Southern California, Center for Computational Study of Macromolecular Structure-Function
US	University of Tennessee at Knoxville
US	University of Texas at Austen, Dept. of Chemical Engineering and NSF STC, Center for Synthesis, Growth, and Analysis of Electronic Materials
US	University of Texas at El Paso, Center of Excellence in Materials Science
US	University of Utah, Dept. of Physics
US	University of Washington, Center of Bioengineering and NSF, STC for Molecular Biotechnology, and School of Medicine, Department of Orthopaedics

US	University of Wisconsin at Madison, Materials Science Center, Institute for Enzyme Research, and Plasma-Aided Manufacturing Engineering Research Center
US	Virginia Commonwealth University, Department of Physics
US	Virginia Polytechnic Institute and State University, NSF STC, Center for High-Performance Polymeric Adhesives and Composites
US	Washington State University, Department of Chemistry, Center for Materials Research
US	Wayne State University, Dept. of Physiology
US	William Marsh Rice University
US	Xerox Corporation, Webster Research Center
US	Xerox PARC
US	Yale University

Bibliography

(Ball, 1993) Philip Ball, "Small Problems," *Nature*, Vol. 362, March 11, 1993.

(Ball, 1994) Philip Ball, *Designing the Molecular World: Chemistry at the Frontier*, Princeton University Press, 1994.

(Binning and Rohrer, 1985) G. Binning and H. Rohrer, "The Scanning Tunneling Microscope," *Scientific American*, Vol. 253, pp. 40–46, 1985.

(Birge, 1992) Robert Birge, "Molecular Electronics," *Nanotechnology: Research and Perspectives*, B. C. Crandall and James Lewis, eds., The MIT Press, 1992.

(Brendley and Steeb, 1993) Keith W. Brendley and Randall Steeb, *Military Applications of Microelectromechanical Systems*, RAND, MR-175-OSD/AF/A, 1993.

(Brumer, 1995) Paul Brumer and Moshe Shapiro, "Laser Control of Chemical Reactions," *Scientific American*, Vol. 272, No. 3, pp. 56–63, March 1995.

(Cahn, 1990) Robert W. Cahn, "Nanostructured Materials," *Nature*, Vol. 348, pp. 389–390, November 29, 1990.

(Carter, 1982) F. L. Carter, ed., *Molecular Electronic Devices*, New York: Dekker, 1982.

(Cichonski, 1994) Thomas J. Cichonski, ed., *Research Centers Directory*, 18th Edition, Vol. 1, Gale Research Inc., Detroit, MI, 1994.

(Conrad, 1992) Michael Conrad, ed., *Computer*, IEEE Computer Society, Special Issue: Molecular Computing, Vol. 25, No. 11, November 1992.

(Crandall and Lewis, 1992) B. C. Crandall and James Lewis eds., *Nanotechnology: Research and Perspectives*, The MIT Press, 1992.

(Dagani, 1995) Ron Dagani, "Improved Organic Transistors, Light-Emitting Devices Developed," *Chemical & Engineering News*, pp. 42–45, April 24, 1995.

(Drexler, 1986) K. Eric Drexler, *Engines of Creation: The Coming Era of Nanotechnology*, Doubleday, New York, 1986.

(Drexler, 1992) K. Eric Drexler, *Nanosystems: Molecular Machinery, Manufacturing, and Computation*, John Wiley & Sons, Inc., 1992.

(Drexler, Peterson, with Pergamit, 1991) K. Drexler, C. Peterson, with G. Pergamit, *Unbounding the Future: The Nanotechnology Revolution*, William Morrow and Company, Inc., 1991.

(Feynman, 1960) Richard Feynman, "There's Plenty of Room at the Bottom," *Engineering and Science* (California Institute of Technology) February 1960.

(Franks, 1991) A. Franks, "Nanotechnology," J. W. Gardner and H. T. Hingle eds., *From Instrumentation to Nanotechnology*, Gordon and Breach Science Publishers, 1991.

(Gopel and Ch. Ziegler, 1992) W. Gopel and Ch. Ziegler, eds., *Nanostructures Based on Molecular Methods*, VCH, 1992.

(Hall, 1993) J. Storrs Hall, "Overview of Nanotechnology," 1993, *Internet Frequently Asked Questions (FAQ)* on sci.nanotech/Usenet Newsgroup. (Adapted from papers by Ralph C. Merkle and K. Eric Drexler.)

(Hameroff et al., 1992) Stuart R. Hameroff, Judith E. Dayhoff, Rafael Lahoz-Beltra, Alexei V. Samsonovich, and Steen Rasmussen, "Models for Molecular Computation: Conformational Automata in the Cytoskeleton," *Computer*, IEEE Computer Society, Special Issue: Molecular Computing, Vol. 25, No. 11, pp. 30–40, November 1992.

(Heinlein, 1940) Robert A. Heinlein, *Waldo & Magic Inc.*, A Del Rey Book, Ballantine Books, 1940.

(Hundley and Gritton, 1994) Richard O. Hundley and Eugene C. Gritton, *Future Technology-Driven Revolutions in Military Operations: Results of a Workshop*, RAND, DB-110-ARPA, 1994.

(Jones, 1995) David E. H. Jones, "Technical Boundless Optimism," *Nature*, Vol. 374, April 27, 1995.

(Kaminuma and Matsumoto, 1991) Tsuguchika Kaminuma and Gen Matsumoto, eds., *Biocomputers*, Chapman and Hall, 1991.

(Kaszynski et al., 1992) Piotr Kaszynski et al., "Toward a Molecular-Size 'Tinkertoy' Construction Set. Preparation of Terminally Functionalized [n] Staffanes from [1.1.1] Propellane, *Journal of the American Chemical Society*, Vol. 114, pp. 601–620, January 15, 1992.

(Kirk and Reed, 1992) Wiley P. Kirk and Mark A. Reed, eds., *Nanostructures and Mesoscopic Systems, Proceedings of the International Symposium*, Santa Fe 1991, Academic Press, Inc., 1992.

(MacDonald, 1992) N. C. MacDonald, "Single Crystal Silicon Nanomechanisms for Scanned-Probe Device Arrays," *IEEE Solid-state Sensor and Actuator Workshop*, Hilton Head Island, South Carolina, June 22–25, 1992.

(McKendree, forthcoming 1995) Thomas L. McKendree, "Planning Scenarios for Space Development," Proceedings of Space Manufacturing X, Barbara Faughnan, ed., American Institute of Aeronautics and Astronautics, forthcoming 1995.

(McKendree, 1993) Thomas L. McKendree, "Educating Systems Engineers for Impending Metatechnology in Product and Process Systems," Proceedings of the Third Annual International Symposium of the National Council on Systems Engineering, James E. McAuley and William H. McCumber, eds., National Council on Systems Engineering, 1993.

(Moffat, 1992) Anne Simon Moffat, "Controlling Chemical Reactions with Laser Light (Bond-Selective Chemistry)," *Science*, Vol. 255, pp. 1643–1644, March 27, 1992.

(OTA, 1991) U.S. Congress, Office of Technology Assessment, *Miniaturization Technologies*, OTA-TCT-514, Washington, DC: U.S. Government Printing Office, November 1991.

(Regis, 1995) Ed Regis, *Nano! Remaking the World Atom by Atom*, Little, Brown and Company, 1995.

(Sarid, 1991) Dror Sarid, *Scanning Force Microscopy*, Oxford University Press, 1991.

(Scanlon and Schultz, 1991) "Recent Advances in Catalytic Antibodies," Philos. Trans. R. Soc. London, Series B, Vol. 332, pp. 157–164 (1991).

(Schneiker, 1989) Conrad Schneiker, "Nanotechnology with Feynman Machines: Scanning Tunneling Engineering and Artificial Life," *Artificial Life*, Christopher G. Langton, ed., Vol. VI, Addison-Wesley, 1989.

(Sligar and Salemme, 1992) Stephen G. Sligar and F. Raymond Salemme, "Protein Engineering for Molecular Electronics," *Current Opinion in Biotechnology*, Vol. 3, 1992, pp. 388–394.

(Taniguchi, 1974) N. Taniguchi, "On the Basic Concept of Nanotechnology," *Proc. Int. Conf. Prod. Eng.*, JSPE, Tokyo, 1974.

(Tomalia, 1995) Donald Tomalia, "Dendrimer Molecules," *Scientific American*, pp. 62–66, May 1995.

(von Neumann, 1966) John Von Neumann, *Theory of Self-Reproducing Automata*, A. W. Burks, ed., University of Illinois Press, 1966.

(Whitehouse and Kawata, 1990) D. J. Whitehouse and K. Kawata, eds., *Nanotechnology: Advances in Nanoscale Physics, Electronics, and Engineering*, Proceedings of the Joint Forum/ERATO Symposium held at Warwick University, August 23–24, 1990, Adam Hilger, Bristol, England, 1990.

(Whitesides et al., 1991) George M. Whitesides, John P. Mathias, and Christopher T. Seto, "Molecular Self-Assembly and Nanochemistry: A Chemical Strategy for the Synthesis of Nanostructures," *Science*, Vol. 254, November 29, 1991, pp. 1312–1319.